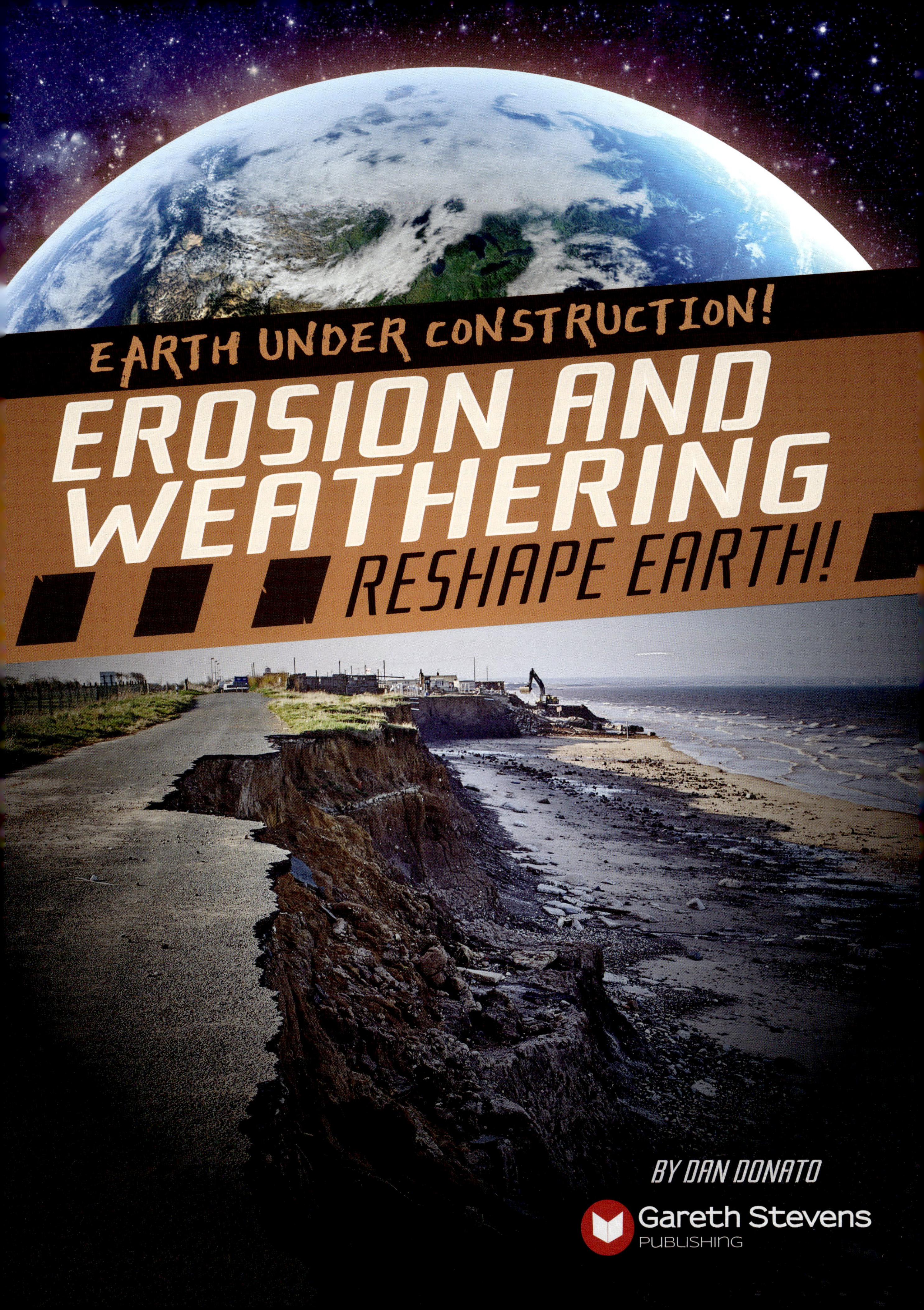
EARTH UNDER CONSTRUCTION!
EROSION AND WEATHERING
RESHAPE EARTH!
BY DAN DONATO
Gareth Stevens
PUBLISHING

Please visit our website, www.garethstevens.com. For a free color catalog of all our high-quality books, call toll free 1-800-542-2595 or fax 1-877-542-2596.

Cataloging-in-Publication Data

Names: Donato, Dan.
Title: Erosion and weathering reshape Earth! / Dan Donato.
Description: New York : Gareth Stevens Publishing, 2021. | Series: Earth under construction! | Includes glossary and index.
Identifiers: ISBN 9781538258262 (pbk.) | ISBN 9781538258286 (library bound) | ISBN 9781538258279 (6 pack)
Subjects: LCSH: Erosion–Juvenile literature. | Weathering–Juvenile literature. | Geochemical cycles–Juvenile literature.
Classification: LCC QE571.D655 2021 | DDC 551.3'02–dc23

First Edition

Published in 2021 by
Gareth Stevens Publishing
111 East 14th Street, Suite 349
New York, NY 10003

Designer: Sarah Liddell
Editor: Kate Mikoley

Photo credits: Cover, p. 1 Matthew J Thomas/Shutterstock.com; space background and earth image used throughout Aphelleon/Shutterstock.com; caution tape used throughout Red sun design/Shutterstock.com; p. 5 (physical) Abdulkhaliq Alshukaili/EyeEm/EyeEm/Getty Images; p. 5 (chemical) Rebecca E Marvil/Photolibrary/Getty Images Plus/Getty Images; p. 5 (biological) Jason Anstett/EyeEm/Getty Images; p. 7 UniversalImagesGroup/Contributor/Universal Imges Group/Getty Images; p. 9 wickerwood/Shutterstock.com; p. 11 Fabrizio Mazzeo/EyeEm/EyeEm/Getty Images; p. 13 Tim Truby/500px/Getty Images; p. 15 marcoisler/RooM/Getty Images; p. 17 MichaelStephenWills/iStock/Getty Images Plus/Getty Images; p. 19 Historical/Contributor/Corbis Historical/Getty Images; p. 21 htu/Moment/Getty Images; p. 23 trabantos/iStock/Getty Images Plus/Getty Images; p. 25 Alan Majchrowicz/Stone/Getty Images; p. 27 Amanda Hall/robertharding/robertharding/Getty Images Plus/Getty Images; p. 29 Harald Sund/Photographer's Choice/Getty Images Plus/Getty Images.

Printed in the United States of America

CPSIA compliance information: Batch #CS20GS: For further information contact Gareth Stevens, New York, New York at 1-800-542-2595.

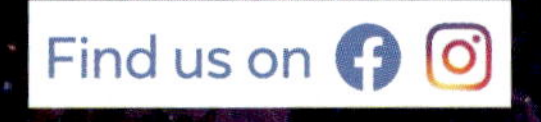

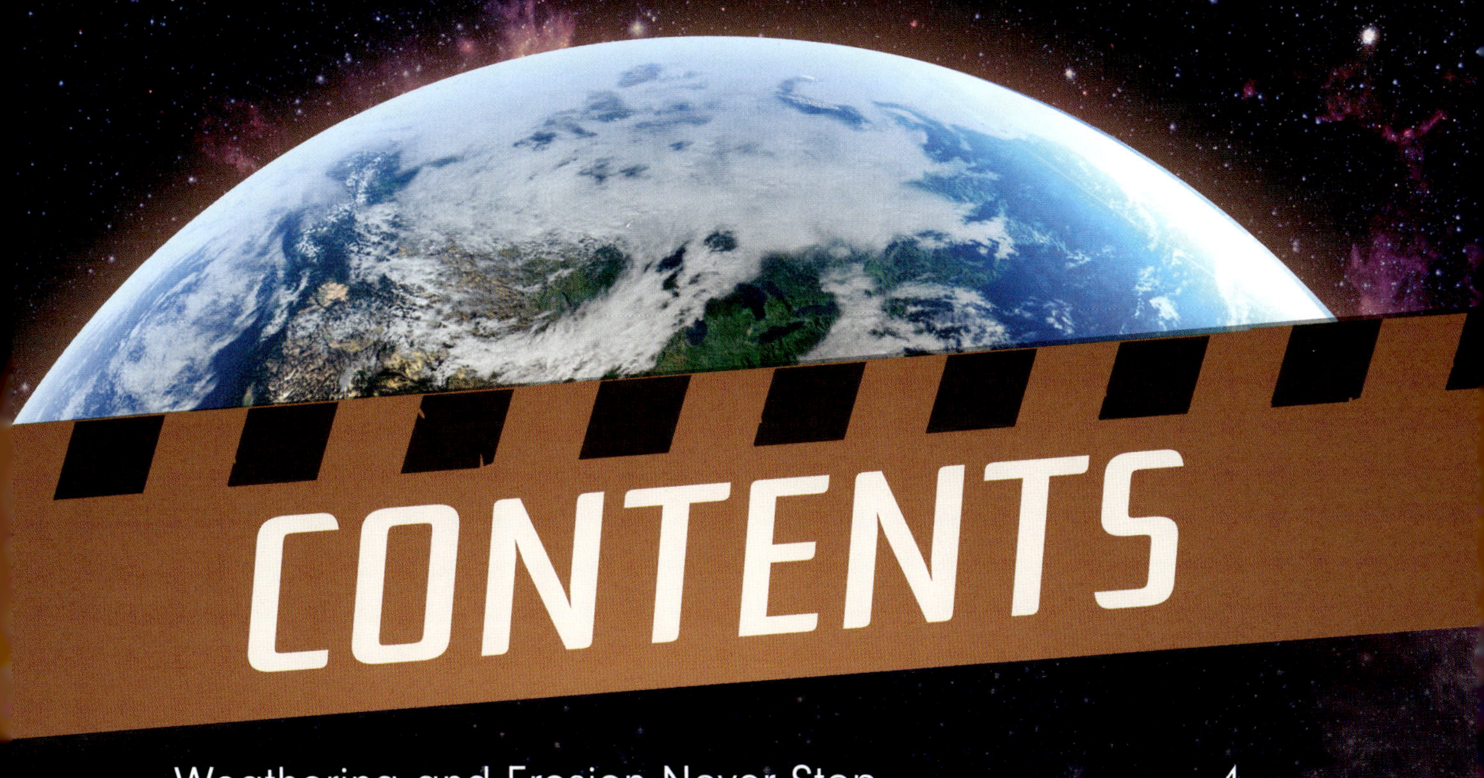

CONTENTS

Words in the glossary appear in **bold** type the first time they are used in the text.

WEATHERING AND EROSION NEVER STOP

A mountain may seem **permanent**, but weathering and erosion never stop. Over millions of years, a mountain can erode so much that there's nothing left! Weathering is when parts of Earth are changed by a range of conditions. Often, this means rocks are broken down into smaller pieces. Erosion is when those smaller pieces are taken away. Water, wind, ice, and chemicals shape the planet through weathering and erosion.

Physical weathering happens when rocks break down from physical processes, such as freezing water causing cracks to expand, or get bigger. Chemical weathering happens when a chemical reaction changes the **molecules** of a rock or other object, causing it to break.

TYPES OF WEATHERING

PHYSICAL
OBJECT BREAKS APART FROM MOVEMENTS THAT OCCUR BECAUSE OF NATURAL CONDITIONS SUCH AS RAIN AND WIND

BIOLOGICAL
OBJECTS WEAKEN OR BREAK BECAUSE OF PLANTS OR ANIMALS

CHEMICAL
OBJECT WEAKENS AND BREAKS BECAUSE OF A CHEMICAL REACTION

FROM CLIFFS TO DIRT WHERE ANIMALS BUILD THEIR HOMES, EROSION AND WEATHERING CAN HAPPEN ANYWHERE!

BIOLOGICAL PROCESSES

Plants and animals—including people—also cause erosion and weathering. When people cut down trees, there are no roots to hold the soil in place, making it more likely to erode. When animals dig holes, pieces of rock that were once protected from certain forms of weathering may be moved to places where they're more exposed. These are examples of **biological** processes of erosion and weathering.

STREAM TO RIVER

It often snows at the tops of mountains. When the snow melts, water runs down the mountain. Over time, this stream of water will weather and erode the ground. When many of these streams flow to the same place, they can form a river.

The water of a river weathers everything it flows over. As it does this, erosion carries the weathered pieces away, making the river grow bigger over time. Small pieces of **sediment** carried by the river are called silt. Silt can be very good for farming because it is very **fertile**. Farmers often plant crops near rivers they know contain a lot of silt.

AS TINY BITS OF ROCK FLOW DOWN A RIVER, THEY HIT AGAINST THE BASE OF THE RIVER, CHIPPING OFF MORE ROCK. THESE PIECES KEEP HITTING EACH OTHER AND BREAKING DOWN UNTIL THEY BECOME SILT.

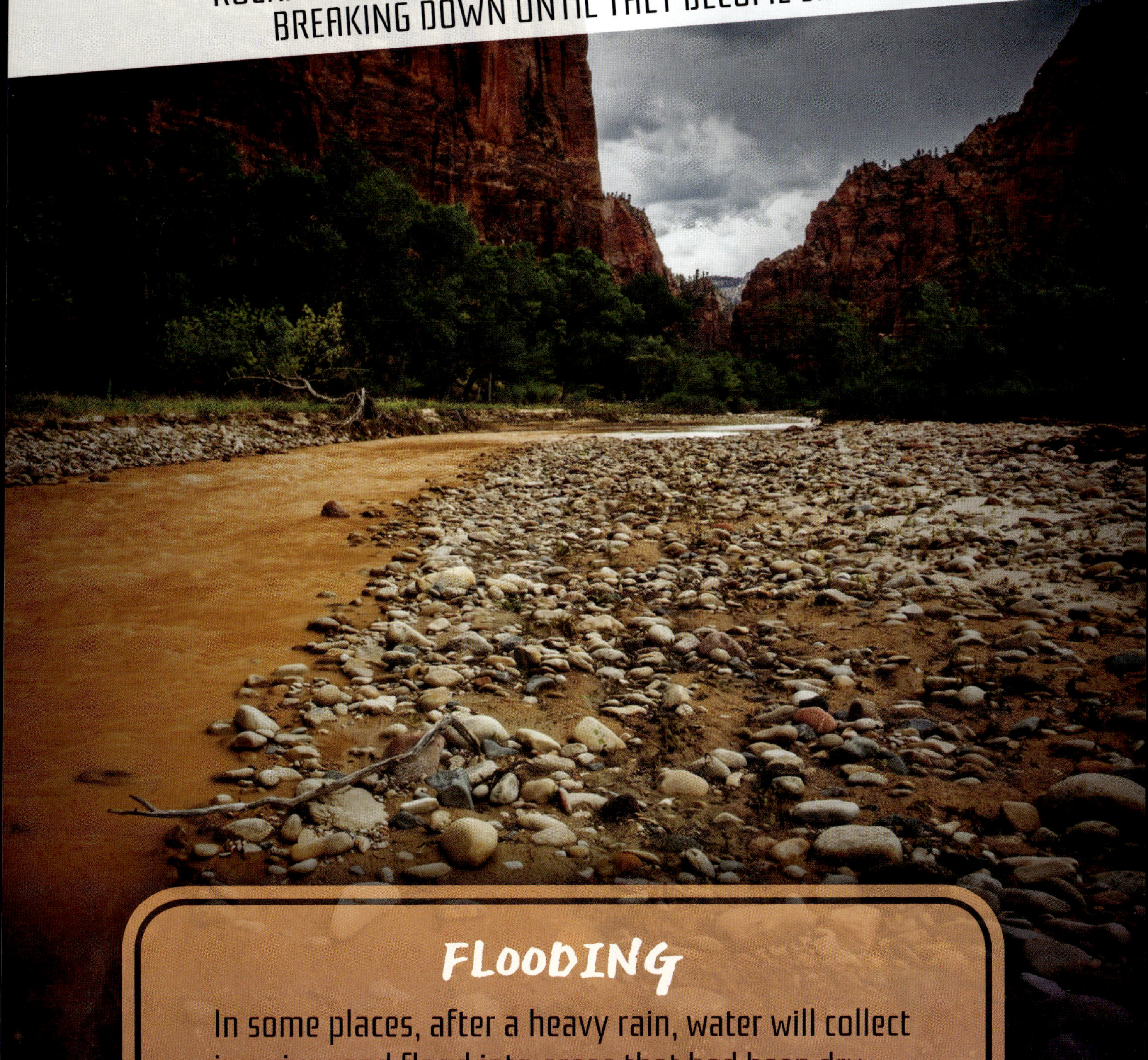

FLOODING

In some places, after a heavy rain, water will collect in a river and flood into areas that had been dry before the rain. Some time after the rain is over, rivers usually return to normal size, leaving the dry areas covered in fertile silt. Farmers often plan around this flooding to grow their crops.

WHY DO RIVERS CURVE?

Erosion happens at different speeds in different places along a river. Dirt erodes faster than rock. A curve is created when something near a river, such as an animal burrow, loosens the dirt, making it wash away more easily. This creates a hole and when water rushes in, it becomes a curve in the river.

The water on the outside of a curve moves faster than the water on the inside. Faster water causes more erosion. Sediment gets **deposited** on the slower, inside part of the curve while more erosion keeps happening on the outside. This makes the curve grow even bigger!

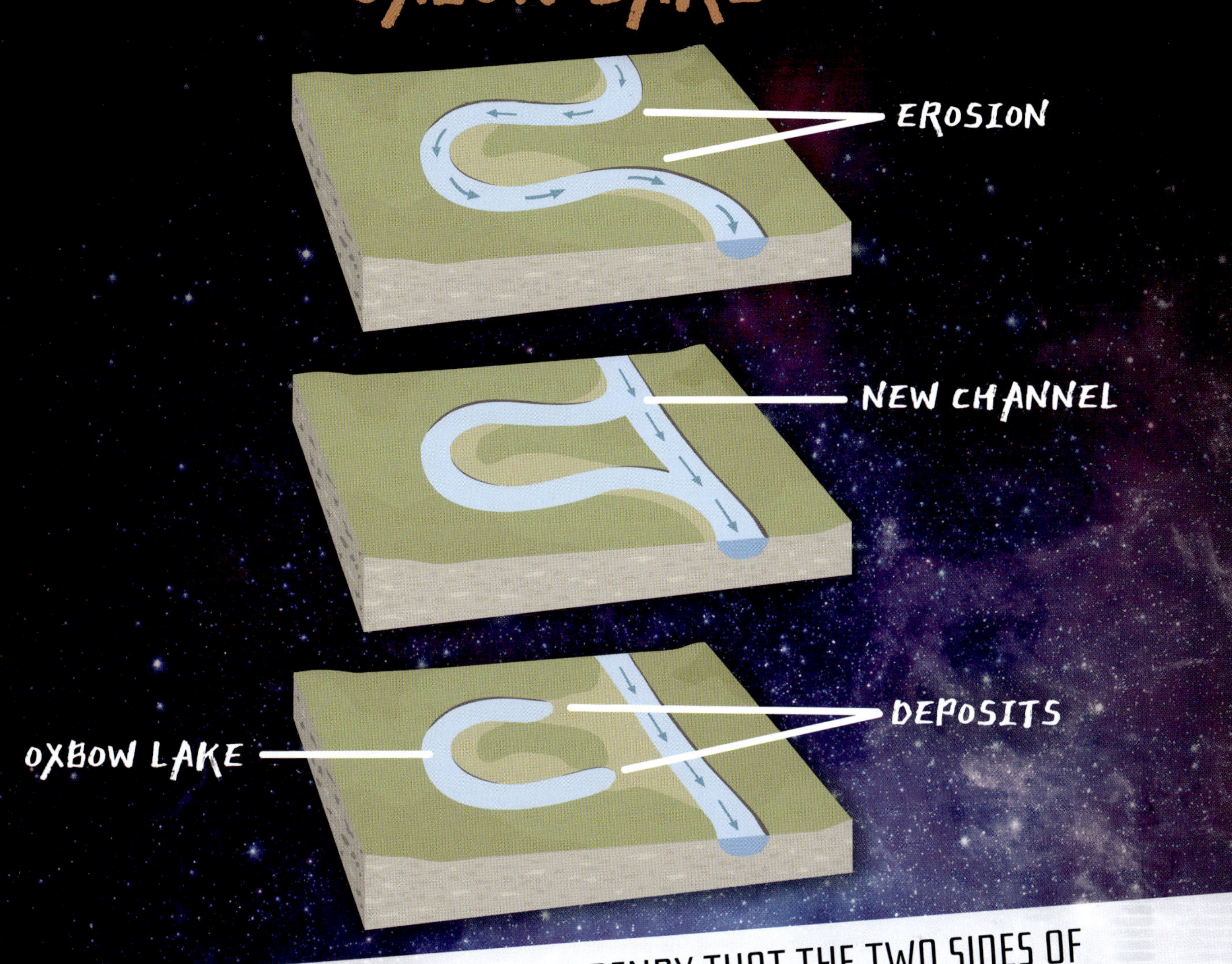

SOMETIMES A RIVER GETS SO BENDY THAT THE TWO SIDES OF A CURVE HIT EACH OTHER, OR THE GAP BETWEEN THE SIDES MAY GET SO SMALL THAT WATER PUSHES OVER IT. THE WATER FORMS A NEW CHANNEL FOLLOWING THE MOST DIRECT PATH. THE BODY OF WATER LEFT BEHIND IS CALLED AN OXBOW LAKE.

WINDING RIVERS

As water quickly moves along the outside of a curve, it gains a force called momentum. This makes it move faster as it whips around the bend and hits the opposite side. This forms a new curve on that side. Over time, this keeps happening. That's why some rivers seem to wind back and forth endlessly.

MOUNTAIN TODAY, CANYON IN A MILLION YEARS

Today, the Grand Canyon stretches for a length of about 277 miles (446 km) in Arizona. At its broadest point, it's about 18 miles (29 km) wide. At its deepest, it goes down for more than 1 mile (1.6 km). But 6 million years ago, the Grand Canyon didn't exist!

We can thank erosion for the creation of this famous formation. Over millions of years, the Grand Canyon was cut away by erosion from the Colorado River. Over time, the water weathered the sides of the cliffs and carried rock away, slowly forming one of the most recognizable land features in the world.

THE ROCK AT THE VERY BOTTOM OF THE GRAND CANYON IS MORE THAN 2.5 BILLION YEARS OLD!

STRIPES OF TIME

If you ever visited the Grand Canyon, you probably noticed the stripes along its sides. These stripes are actually layers of different types of rocks from different periods of time. The layers at the bottom of the canyon are the oldest. By studying the layers, scientists can learn about different times in Earth's history.

SNEAKY SAND

Sand comes from rocks that have been broken down by weathering and erosion for many years. As weathering happens, erosion often carries the sand through rivers or streams, and often, into the ocean. The ocean's tide carries and deposits sand as its moves. This is why we have beaches!

In sandy areas, like beaches or deserts, wind often continues the cycle of erosion. When the wind blows, it picks up bits of sand, eroding the surface the sand once covered. As the sand moves with the wind, it hits anything in its path, causing even more erosion.

BLACK SAND IS USUALLY FOUND ON BEACHES WITH **VOLCANOES** NEARBY, SUCH AS THIS ONE IN HAWAII.

AT THE BEACH

Some beaches have really white sand. Others have tan or brownish sand. Some even have black sand. The different colors are results of the different rocks that make up each kind of sand. For example, black sand comes from dark rocks and minerals, such as a hardened **lava** called basalt.

MORE OCEAN EROSION

Ocean waves cause erosion by smashing against the coast. The waves weather and erode the coastline, carrying away small pieces of land such as pebbles and sand. These pieces then help continue erosion by wearing and grinding against the remaining shore. This is why there are often cliffs along the shore of the ocean.

Salt from the ocean also weathers rock. Ocean waves can leave behind salt water that gets trapped in cracks of the rock. The water evaporates, or turns to a gas, and leaves salt crystals behind. As the salt crystals grow, they push against the rock from the inside causing it to break. This makes it a physical weathering process.

SEA STACKS ARE PARTS OF A CLIFF THAT HAVE BEEN SEPARATED BY EROSION OVER MANY YEARS. THEY LOOK LIKE A STACK OF ROCKS STANDING ALONE IN THE WATER.

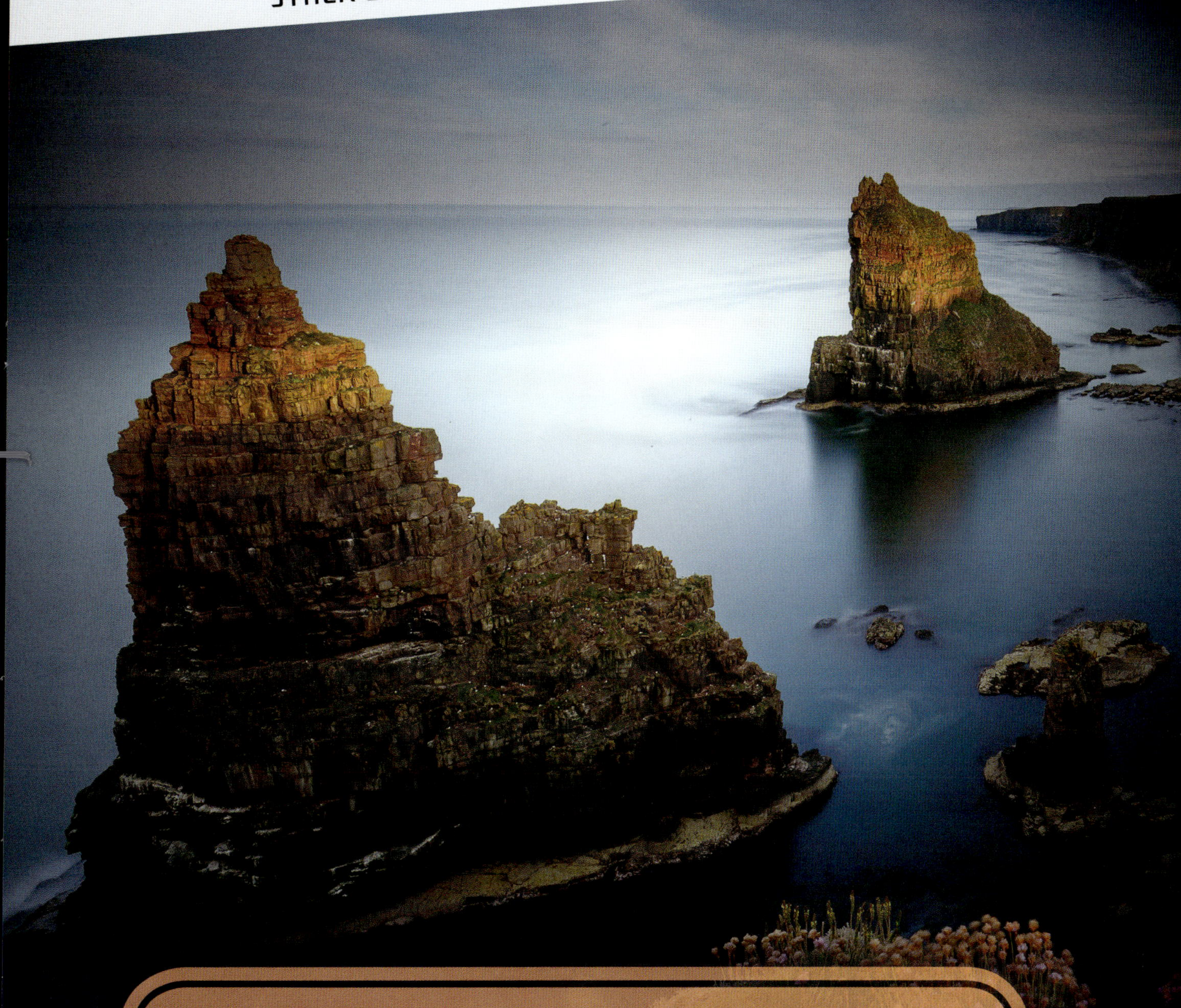

SANDCASTLES AND REAL CASTLES

One way to think about erosion is to think about a sandcastle being washed away by waves. A real castle made of stone is much stronger than a sandcastle because stone is stronger than sand, but if it were hit by waves, over time, it would erode too!

CAN A FARM BLOW AWAY?

To protect their crops, farmers need to know about soil erosion. Too much rain can carry away the soil used to grow crops. To stop this from happening, farmers grow plants that cover the soil and help keep it in place. Called cover crops, these plants have strong roots that grip into the soil.

Farms are also at risk from wind. Wind erosion can blow seeds and soil away. Wind can also carry rocks and dirt and leave them behind on the farm, sometimes burying crops. This makes the land harder to farm. To block the wind, farmers plant tall trees and bushes in lines, called windbreaks.

SOME TREES USED FOR WINDBREAKS GROW QUICKLY. PARTS OF THESE TREES CAN BE CUT OFF. THE WOOD CAN BE SOLD FOR EXTRA MONEY OR USED FOR BUILDING.

MOUNTAIN FARMS

Some mountains or hills have great soil for farming. It might seem hard to find a way to farm these areas without soil being washed away by rain, but terrace farms make it possible. Sloped land is formed into steps. When rain falls, channels bring the soil to the next step down, rather than washing it away.

THE DUST BOWL

During the 1930s, there was a period of very little rain in parts of the United States' Great Plains, which became known as the Dust Bowl. Soil was dry and easy for the wind to blow away. As wind blew, it eroded the soil, causing "black blizzards," which blocked sunlight. Soil fell in large piles called drifts.

Many people lost their farms and had to leave the area because of the Dust Bowl. Eventually, the federal government helped plant windbreaks and wind erosion slowed. By the 1940s, the area was mostly back to normal, but around 2.5 million people had already left their homes behind.

THE DUST BOWL AFFECTED PARTS OF COLORADO, KANSAS, TEXAS, OKLAHOMA, NEW MEXICO, AND NEBRASKA.

FROM DUST TO DEPRESSION

At the same time the Dust Bowl was happening, the entire United States was going through the Great Depression. This was a period of widespread **poverty** when many people could not find jobs. The Dust Bowl made this even harder. Many farmers moved to California, only to find different challenges caused by the Great Depression.

GIANT MOVING ICE

Glaciers are giant bodies of ice that move slowly over an area of land. As a glacier moves, it scrapes at the land underneath it. This is called abrasion. The scraped up rock or dirt is then carried away through erosion, leaving behind stripe-like scratches or marks.

As glaciers move, melting ice can find its way into cracks in the rock below. As this water refreezes, it expands. This makes the cracks in the rock bigger and makes it easier for large pieces of the rock to break off. The glacier will then carry these pieces away. This is called plucking.

AFTER A GLACIER CARRIES AWAY BITS OF ROCK, IT MAY DEPOSIT THEM IN A NEW AREA. IF ENOUGH ROCK GETS LEFT BEHIND, IT CAN MAKE A NEW LANDFORM.

MAKING A GLACIER

A glacier forms when more snow gathers than can melt away. Over time, the snow at the bottom freezes together. As new snow falls, it freezes onto the forming glacier. When enough snow gathers, the pressure begins to melt the bottom layer causing the glacier to move—and causing erosion!

WEATHERING IN CAVES

Rain mixes with carbon dioxide in the air. As it mixes with the soil, the water can become acidic. It causes a chemical reaction that can break down certain rocks it touches. This is how many caves form. It's a process of chemical weathering.

Inside some caves, you might see tall pointed rocks, called stalagmites or stalactites. Stalagmites grow up from the floor and stalactites hang down from the ceiling, like icicles. As water coming into the cave weathers the rock away, it takes in a chemical called calcium carbonate. Stalagmites and stalactites are the results of this water dripping in the cave.

MANY CAVES FORM FROM A KIND OF ROCK CALLED LIMESTONE. LIMESTONE HAS A LOT OF CALCIUM CARBONATE, SO STALAGMITES AND STALACTITES ARE COMMON IN THESE CAVES.

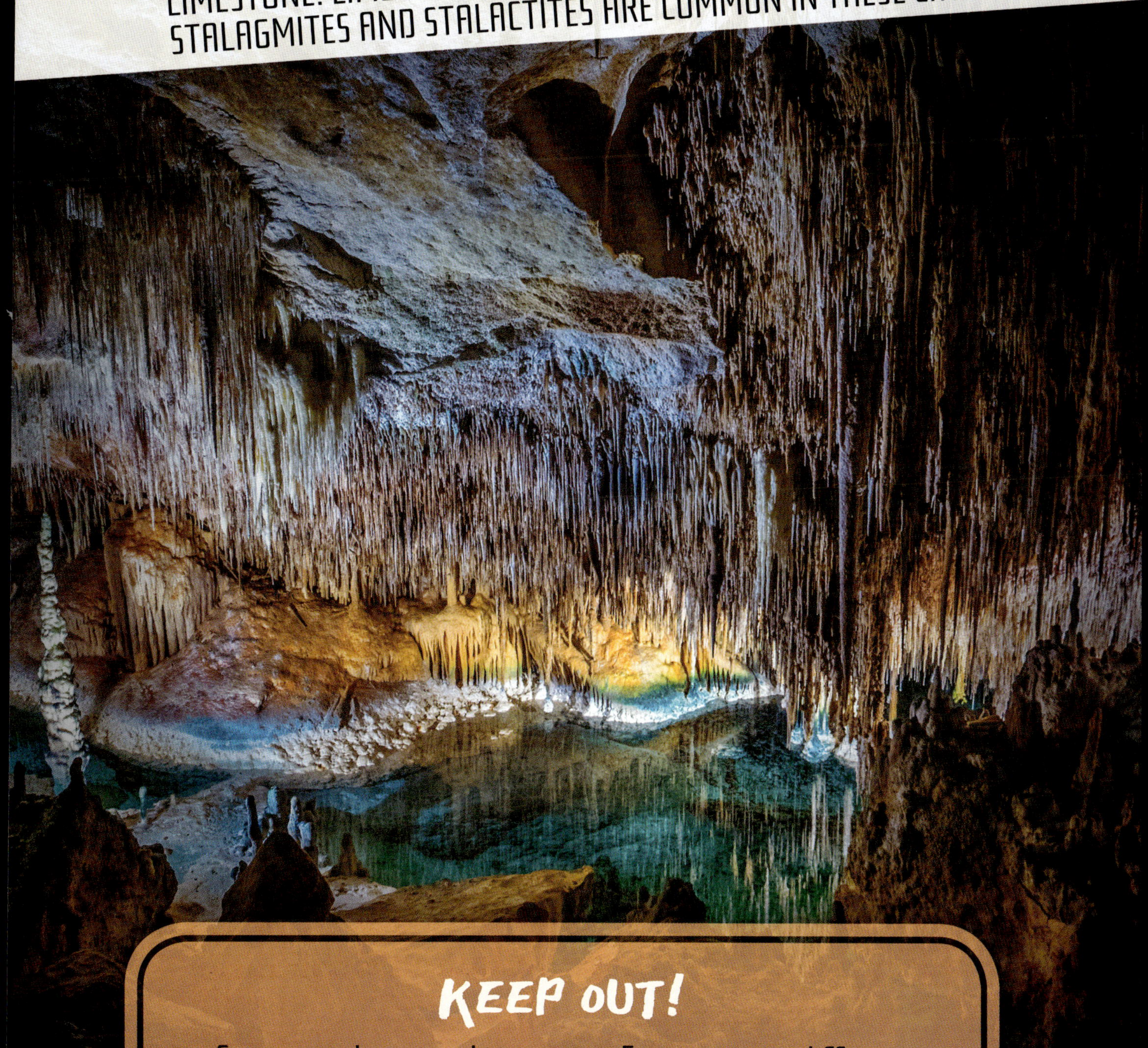

KEEP OUT!

Caves can be very dangerous. Every cave is different and sunlight doesn't reach the inside, so it can be easy to get lost. Erosion of the land above the cave can also cause a cave to collapse. Experts who explore caves are known as spelunkers. They make maps to stay safe in the caves they explore.

BLINDING FACTS ABOUT THE SUN

Most solid objects expand, or get bigger, when they're heated. This is because their atoms have more energy when heated and move farther apart. In particularly warm areas, this can be a cause of weathering. During the day, the side of a rock facing the sun gets hot and expands. At night, this part of the rock cools down and goes back to its slightly smaller size.

These changes can cause the rock to get weaker, and eventually crack or break off completely. In sloped areas, the weathered pieces may simply fall away. In flatter areas, they may be eroded over time with the help of wind or water.

HEAT FROM THE SUN CAN CAUSE LARGE CRACKS IN SOIL. THIS IS COMMON WHEN THE SOIL HAS A LOT OF CLAY. CLAY TAKES IN MOISTURE DURING WET PERIODS, AND DRY PERIODS CAUSE TINY PARTS OF CLAY TO PULL TIGHTLY TOGETHER.

SUN FADE

Have you ever left a towel or toy outside for a long time and noticed its color changed? This is a kind of weathering too. Sunlight reacts with **dyes** used to color many objects. The reaction breaks the chemical bonds of the dye. As the bonds break they become weaker, making the color fade.

WHY IS THE STATUE OF LIBERTY GREEN?

We know that weathering can reshape Earth, but it can also reshape man-made objects. A kind of chemical weathering called oxidation happens when an object has a chemical reaction resulting in the loss of electrons.

The Statue of Liberty is made out of copper. It was originally a brownish color like a penny. Today, the statue is green because of oxidation. The moisture in the air caused a chemical reaction that weathered the statue, resulting in a layer of green matter. Though it might seem like a copper object turning green is rotting, the opposite is actually happening. The green matter protects the copper!

FRANCE GIFTED THE STATUE OF LIBERTY TO THE UNITED STATES. IT WAS PUT UP IN 1886. BY 1920, IT HAD TURNED COMPLETELY GREEN.

A LITTLE RUSTY

Oxidation is also the process that causes certain objects made of metal to rust. Rocks can also get rusty if they have iron in them. This is why some rocks are a reddish color. The oxidation weakens the rock, which can cause it to break or crumble over time.

WEATHERING AND EROSION ARE EVERYWHERE

Weathering and erosion are changing rocks and landforms all the time. These changes are often small and can't be seen. Over millions of years, however, these processes can completely change the way parts of Earth look.

Since weathering and erosion can happen slowly and are always going on, it's easy for people not to see these changes. But with knowledge of these processes, we can figure out how the planet around us is going to change. When farmers learn about erosion, they learn how to best protect their crops. Understanding erosion can also help us prevent **disasters**, like the Dust Bowl, from happening again.

GEOLOGY IS THE SCIENCE THAT STUDIES THE HISTORY OF EARTH AND ITS LIFE AS RECORDED IN ROCKS. A SCIENTIST THAT STUDIES THE EROSION OF THE EARTH AND ROCKS IS A GEOLOGIST.

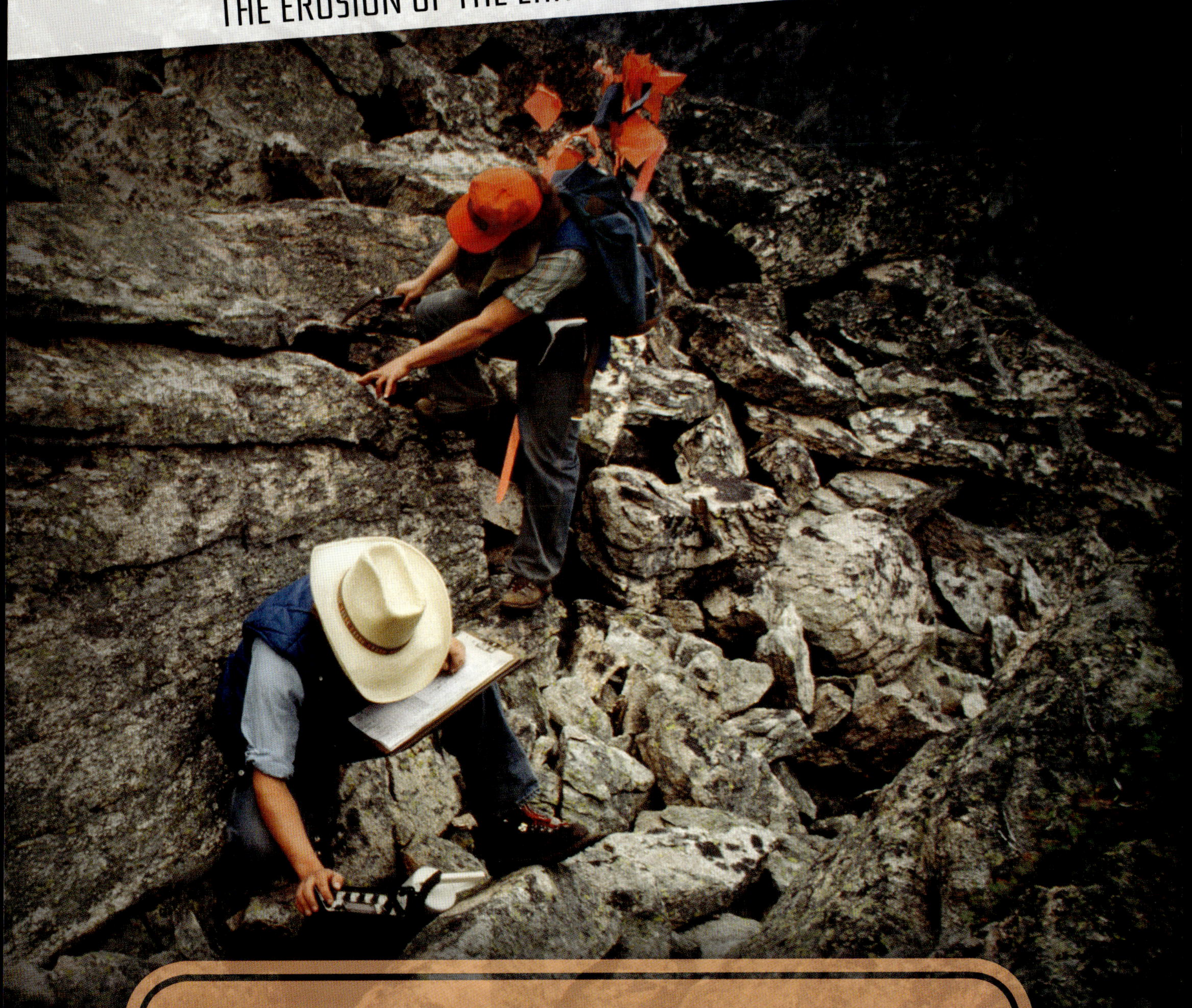

EARTH DETECTIVE

Studying erosion can give you a glimpse into the past. If you see a tall rock in the water, you can bet there was probably once a cliff attached to it. When you see a small lake near a river, you now know it might have once been part of that river. Can you imagine what it might have looked like?

GLOSSARY

biological: having to do with life and living things

deposit: to let fall or sink

disaster: an event that causes much suffering or loss

dye: a substance used to add color to a product such as clothing

fertile: able to support the growth of many plants

lava: melted rock from a volcano

molecule: a very small piece of matter

permanent: unable to be removed

poverty: the state of being poor

sediment: matter, such as stones and sand, that is carried onto land or into the water by wind, water, or land movement

volcano: an opening in a planet's surface through which hot, liquid rock sometimes flows

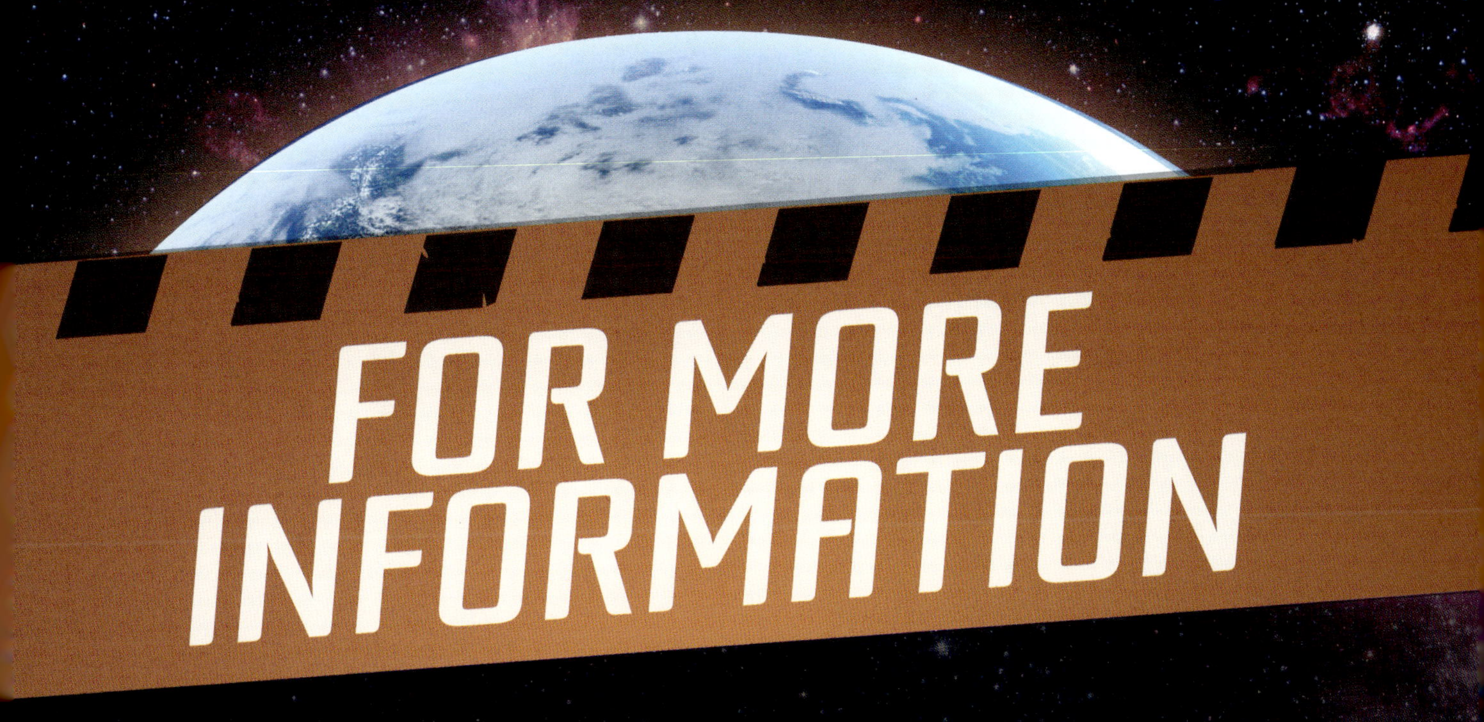

FOR MORE INFORMATION

BOOKS

Machajewski, Sarah. *Storms, Floods, and Erosion.* New York, NY: PowerKids Press, 2019.

McAneney, Caitie. *Weathering and Erosion.* New York, NY: Britannica Educational Publishing, 2018.

Wilson, Steve. *Erosion and Sediments.* New York, NY: PowerKids Press, 2017.

WEBSITES

Erosion
www.ducksters.com/science/earth_science/erosion.php
Head to this page to learn more about the process of erosion.

How Does an Oxbow Lake Form?
www.dkfindout.com/us/earth/rivers/how-does-an-oxbow-lake-form/
Learn more about oxbow lakes on this website.

Weathering
www.nationalgeographic.org/encyclopedia/weathering/
Find out even more about weathering here.

Publisher's note to educators and parents: Our editors have carefully reviewed these websites to ensure that they are suitable for students. Many websites change frequently, however, and we cannot guarantee that a site's future contents will continue to meet our high standards of quality and educational value. Be advised that students should be closely supervised whenever they access the internet.

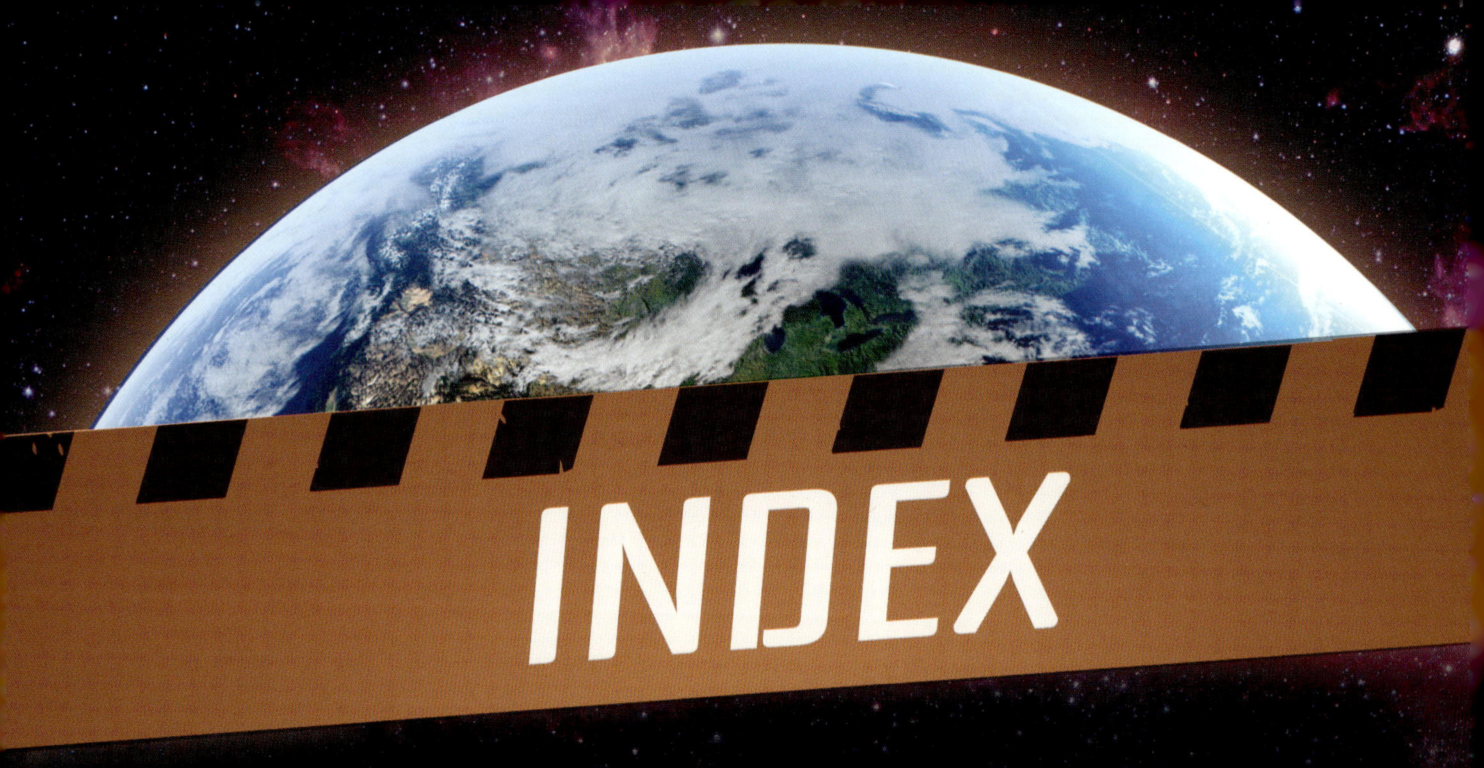

INDEX